D1796423

Pollution

NATIONAL SOCIETY FOR CLEAN AIR

To Lynne and Paul

Printed and Published in Great Britain 1983 by
The National Society for Clean Air
136 North Street, Brighton
East Sussex BN1 1RG
ISBN 0 903474 18 2
Typeset in Univers by Trident Typesetting

CONTENTS

THE NATIONAL SOCIETY FOR CLEAN AIR is a non-governmental organisation and a registered charity, its origins dating back to 1899 (the Coal Smoke Abatement Society). The NSCA exists to secure environmental improvement by promoting clean air through the reduction of air pollution, noise and other contaminants, while having due regard for other aspects of the environment. It does this by advancing the investigation, consideration and discussion of all forms of pollution in order to achieve its reduction or prevention. The NSCA receives no government grant, and is funded by members' subscriptions, donations and monies raised by activities. Public education is an important aspect of the Society's work, accomplished through conferences and other meetings, exhibitions and publications. The Society has its own press and publishes Conferences and Workshop proceedings, a quarterly journal, an annual handbook, books, reports and educational material. A full list of publications is available from the NSCA's Brighton office, which also houses the Society's library and information department.

INTRODUCTION

Pollution, whether of air, water or land, is certainly not a new problem. Poisonous substances have been released into the environment for many years in all industrial countries and areas of dense human habitation; smoke-filled air, dirty rivers and contaminated land have become familiar features of the landscape. What is changing, however, is the attitude of the public.

It has become increasingly clear in recent years that if pollution goes unchecked it could result in serious and possibly irreversible damage to this planet. Of crucial importance is the need to understand the problems; such understanding brings with it the knowledge which will allow control of the adverse effects on human health and the environment. The search continues to identify and control new problems and find solutions to old ones. Provided that man can develop and practise a proper respect for the world, there is no reason to suppose that he will not be able to continue to enjoy life on this planet for many generations.

WHAT IS POLLUTION?

Before beginning to discuss different aspects of the subject, it is important to consider exactly what is meant by "pollution". Used in its broadest sense, the term refers to the effect on quality of the environment brought about by chemical, physical or biological change. Since the basic state of the environment is hard to define, it is not always clear where the word should be applied. An interesting question is whether the changes brought about by nature (for example the effects of volcanic emissions) should be regarded as pollution. In this publication a more limited definition has been applied, restricted to the consequences of human activity.

The problem which mankind faces is to find a balance between the benefits of a rising standard of living, and its costs in terms of damage to the environment and the quality of life. In the past, the danger of polluting air, water and land was not fully recognised, but now there is no doubt that it is a matter of great concern. Equally, many people have reached the limit of their tolerance to noise and demand that they should be protected, especially from the noise of aircraft and traffic. Freedom from such nuisance is now regarded by most people as a basic right. This booklet examines many of the important aspects of the subject and suggests further areas of interest.

AIR POLLUTION

Air is essential to life; its natural quality must be maintained in order to safeguard man's health and well-being and to protect his surroundings. Air is polluted when the presence of a foreign substance or a variation in its components is liable to have a harmful effect or cause nuisance. Brief information on a range of air pollutants is given in the following paragraphs.

Smoke and Sulphur Dioxide

The first recognised and most widely researched pollutants are the fine particles suspended in the atmosphere (called smoke for convenience) and sulphur dioxide (a colourless gas with a choking taste). Smoke is produced by the incomplete combustion of carbon-based fuel, and sulphur dioxide by the burning of sulphur — a natural constituent of both coal and oil. For as long as coal has been used in towns and cities it has caused problems. Its use was prohibited in London in 1273 because of its prejudicial effects on health. Subsequently, various attempts were made to abolish its use permanently, without success. One energetic campaigner was John Evelyn who wrote to Charles II in 1661 on the problem of smoke in London. (The reprint of this tract, *Fumifugium,* is available from the Society). His recommendations went unheeded and the air became progressively dirtier.

As more and more people crowded into towns, air quality deteriorated. Although coal was at first only used by the wealthy for heating, the industrial revolution, with the development of the steam engine, brought about much wider utilisation of coal to provide power. Demand for large work forces near to the factories increased overcrowding. The spread of the canal and railway networks reduced transportation costs, making coal available to an ever increasing population for industrial, commercial and domestic use. Effects of this change became obvious.

Large centres of population, such as London, Birmingham and Manchester, were affected by thick yellow fog during most winters. These fogs were caused by a build-up of smoke in still, cold air. What was later to become known as smog was a regular feature of Victorian life. Charles Dickens frequently refers to dense impenetrable fogs when describing conditions in London during the mid 19th century. Still little was done to control the problem, in spite of the efforts of campaigners who pointed out the dangers of smogs and the means of controlling them. It was not until 1952 that there was an incident of sufficient magnitude to shock the nation. During the early days of December of that year a particularly bad fog settled over London. Traffic ground to a halt, London airport was closed and many sufferers from bronchitis and heart disease became ill.

Illness was followed all too often by death. Over a period of a few days some 4,000 additional deaths were recorded. At last action was widely demanded and after four years of research, deliberation and debate, an Act of Parliament, designed to clean the atmosphere, was passed. The Clean Air Act of 1956 was the first step in a national campaign of positive action. Its major powers included strict limitations on industrial smoke emissions, the ability to stop coal being burnt on open fires in selected areas, and prevention of other forms of smoke nuisance.

Middlesbrough, before and after smoke control. *(Photographs courtesy of Dept. of Planning, Cleveland County Council.)*

At first gradually, then with increasing pace, smoke control areas were introduced in most parts of the country where problems were serious. Grants were given to help people convert their fireplaces to enable them to burn smokeless fuel. Although some people happily changed to specially manufactured smokeless fuels many more chose the convenience of gas or electric heating.

Industry and commerce were required to sort out their problems at their own expense. The smoking chimney is a sign of fuel inefficiency. Many firms, at that time, chose to switch from coal to oil firing, which lent itself to a higher degree of automatic control. When North Sea gas became widely available, its low cost and freedom from dirt attracted many new customers. Its use did much to reduce smoke levels still further.

Having recognised a problem, it was decided that local authorities should measure levels of air pollution in the atmosphere. Monitoring stations were set up throughout the country and the results from these sites have shown a great reduction in pollution levels over some 20 years. The apparatus used to measure air pollution is relatively simple, and is described in the box overpage.

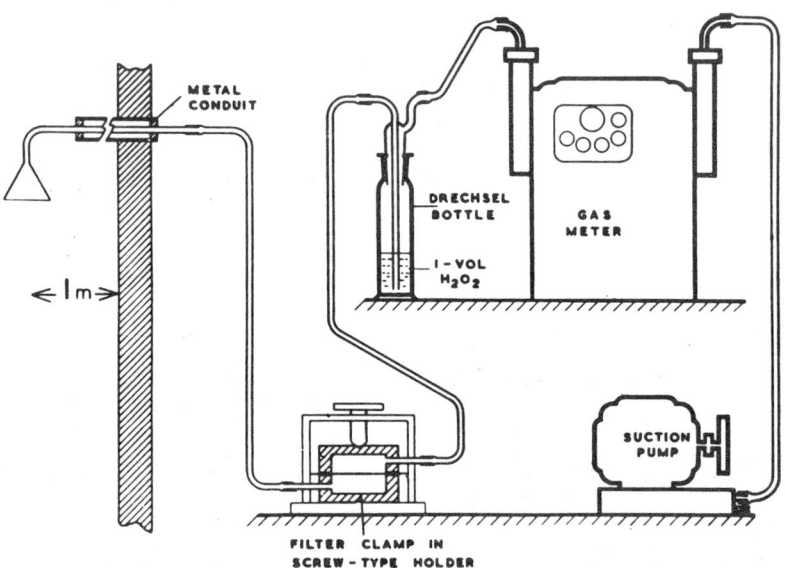

Fig. 1. Schematic arrangement of standard daily smoke and sulphur dioxide sampling apparatus *(Source: Warren Spring Laboratory, National Survey of Smoke and Sulphur Dioxide Instruction Manual.)*

Air is collected from the area of interest via an inverted funnel which is used to prevent rain water being drawn into the equipment. The air is first passed through a sheet of white filter paper, then a bottle containing a dilute solution of hydrogen peroxide. Smoke concentration is measured by assessing the darkness of the stain on the paper; sulphur dioxide, by the change in the acidity of the solution. Because a gas meter records how much air has been sampled each day, it is possible to assess the concentration of both of these pollutants in the atmosphere. Each day the filter paper and the solution are replaced in order to provide results which show changes in pollution concentration. Apparatus has been developed which contains sets of filter papers and bubbling bottles, controlled by an automatic valve, so that stations may operate for a week without requiring servicing.

It should be possible to obtain results of environmental monitoring from the local environmental health department. Relationships between air pollution levels and weather conditions can be obtained by studying such data. This is an ideal subject for a small project.

Fig 2 shows the reduction in urban smoke levels from the early 60s to the 1980s.

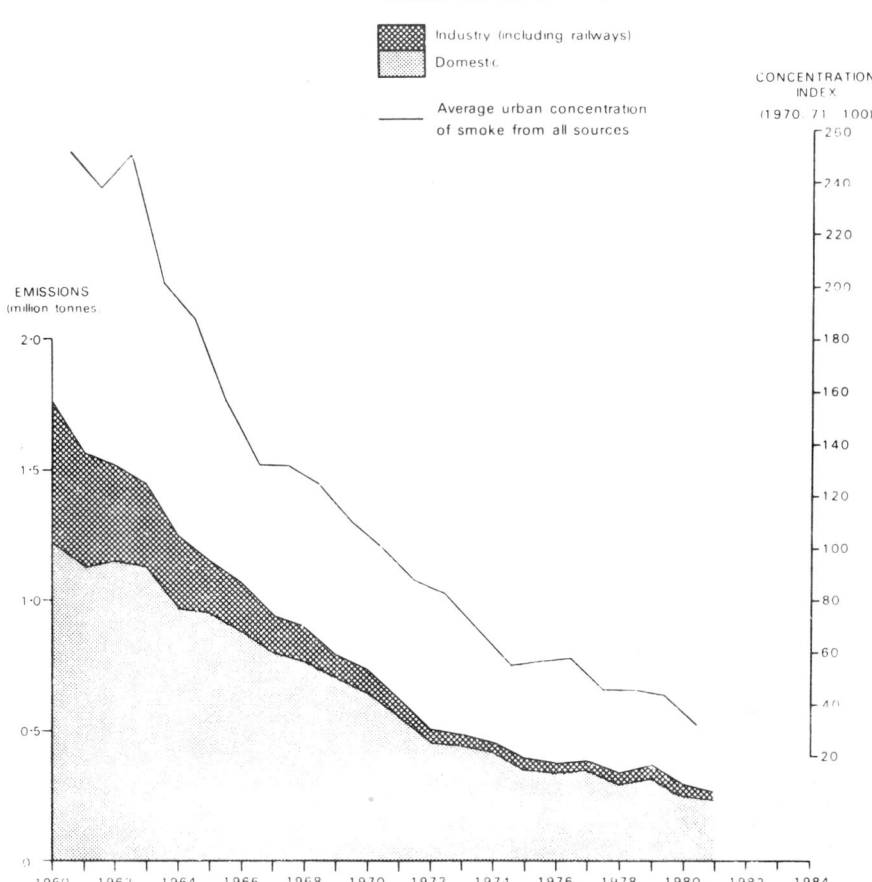

Fig. 2. Smoke: emissions from coal combustion and average urban concentration *(Source: DOE Digest of Environmental Pollution and Water Statistics, No. 5 1982.)*

Reductions in levels of sulphur dioxide have not been so dramatic. Efficient combustion reduces smoke, but it does not affect the level of sulphur dioxide released from the chimney.   Nevertheless, average concentrations of SO_2 in urban areas fell by over 60% between 1960 and 1980 (see Fig. 3), although total emissions of sulphur dioxide only fell by about 20-25%. The improvement in ground level concentrations of sulphur dioxide has been achieved by the switchover to sulphur-free natural gas in industry, mentioned earlier, the use of gas or electricity for heating, and the special control of large scale users of coal or oil.

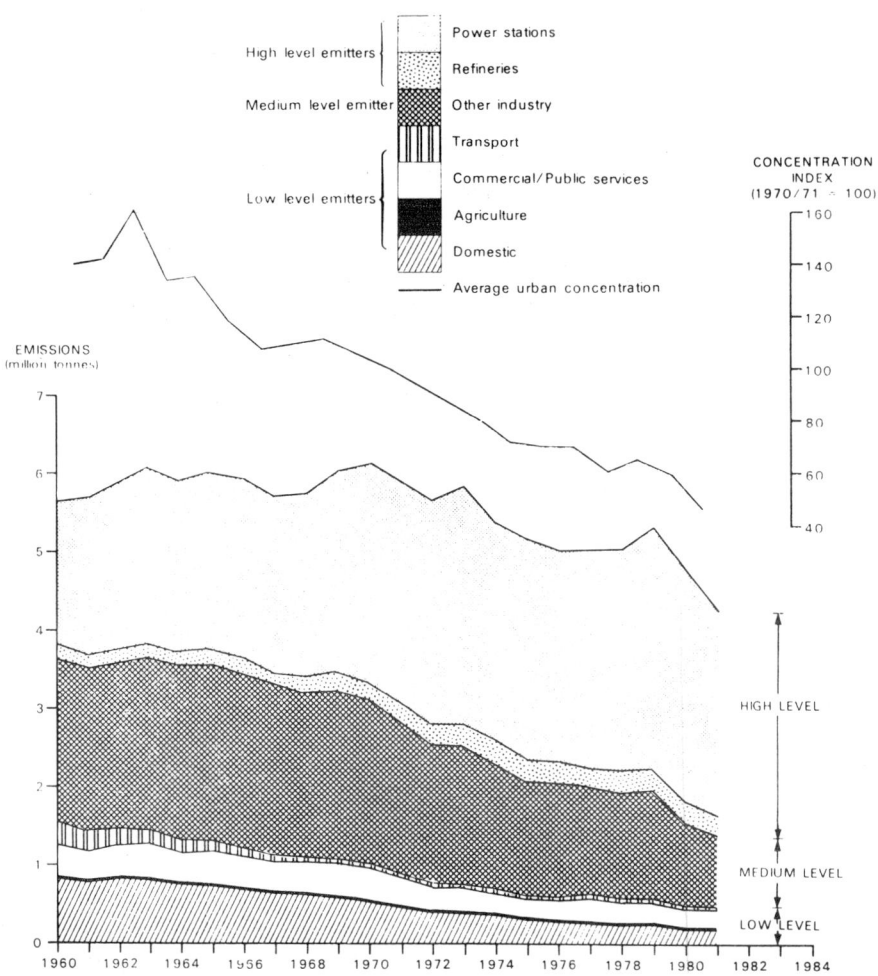

Fig. 3 Sulphur dioxide: emissions from fuel combustion and average urban concentrations *(Source: DOE Digest No 5, 1982.)*

Power stations, refineries and other large industrial establishments are required to build chimneys tall enough to ensure that emissions will be well dispersed and will not create unacceptable ground level conditions. The control of air pollution from these and other special classes of industry is the responsibility of a central government inspectorate, H.M. Industrial Air Pollution Inspectorate (formerly H.M. Alkali and Clean Air Inspectorate). The use of tall chimneys is an element of the 'best practicable means' control approach which is embodied in the Alkali Acts and the Health and Safety at Work Act 1974.

Although it is not a required element of control, attempts have been made in the UK to remove sulphur dioxide before gases are discharged to atmosphere.

The most successful was at Bankside power station, where the gases were 'scrubbed' with water from the River Thames, achieving better than 95% reduction in emissions.

The disadvantages of this method were high operating cost, poor dispersion of the cold wet plume, and damaging effects on the Thames. This scrubbing system no longer operates.

While the situation in the UK has improved, there is increasing concern about the long range impact of sulphur and nitrogen emissions. There have been strong claims that emissions transported across the North Sea by windstreams high in the atmosphere are causing problems in Scandinavia, and there is increasing international effort to establish the effects of acid deposition and suitable methods of control.

There are still parts of the country where large amounts of coal are burnt domestically, particularly in mining areas. All Members of the European Community have agreed on the need to achieve minimum air quality standards for smoke and sulphur dioxide. A comprehensive programme is being drawn up, designed to improve 'black spots' throughout the Community. It should be implemented by 1993 at the latest.

Changes in the availability of gas have brought about modifications in the fuel policy of the United Kingdom. The UK has great reserves of coal and in coming years industry will be actively encouraged to make greater use of them. This does not mean that the country will return to a Dickensian environment. Traditional techniques of coal firing have been improved and new boilers are being designed which are more efficient. There is no reason why smoke should be produced from any new industrial steam-raising plant. One new process, called a fluidised bed, burns coal in suspension. In this type of unit, it is possible to add limestone to the coal as it burns, so that a chemical reaction takes place, reducing the quantity of sulphur dioxide discharged to the atmosphere.

Most old towns have plenty of evidence of the grime and dirt which is encrusted on stone buildings. These are gradually being cleaned, but stone can also be seen to have been eaten away by acid gases. This evidence can be photographed as a record of the effects of air pollution. Look particularly for buildings and monuments made of limestone — although sulphur oxides also erode concrete and cement, and promote corrosion of almost all metals other than gold and the noble metals.

Simple plants, such as lichens, are sensitive to pollution levels, and maps of local pollution can be drawn up by identifying distribution of different species. Full details of this technique are described in: *Air Pollution and Lichens. Edited Ferry, Baddeley and Hawksworth, Pub. Athlone Press.*

Chimneys are not the only source of smoke. At one time, many industrialists chose to rid themselves of waste by burning all of their rubbish on open bonfires. Major restrictions were therefore introduced in the second Clean Air Act, passed by Parliament in 1968.

Individuals, too, create their share of pollution. Garden bonfires cause nuisance to neighbours and fill the air with smoke. Such action is inexcusable and unnecessary. Most garden waste can be composted to the benefit of the soil and community. Perhaps the only bonfires which have to be accepted are the annual celebrations on the 5th November!

Great progress has been made in cleaning the air of the 'traditional' pollutants. The task has not been completed, but already the benefits are considerable.

Sulphates

Although, as indicated above, air pollution has traditionally been measured in terms of sulphur dioxide and smoke, interest is growing in the significance to health of fine particles of sulphate salts in the atmosphere. Recent studies in America have shown that there is a stronger association between a combination of sulphate and suspended particulates than there is between sulphur dioxide and smoke. If the levels of sulphate and suspended particulates were to be halved, it is claimed that life expectancy would be increased by an average of about 9 months.

Nitrogen oxides

Of all the oxides of nitrogen, the significant ones in the context of air pollution are nitric oxide (NO) and nitrogen dioxide (NO_2). Both NO and NO_2 are found in combustion gases; NO predominates as its formation is favoured at high temperatures. Some of the colourless NO oxidises in the atmosphere to NO_2, a reddish-brown gas.

The major man-made source of nitrogen oxides is the combustion of fossil fuel (in power generation, heating plants, road vehicles). Some industrial non-combustion processes are significant local sources of nitrogen oxides (nitric acid plants, electroplating and explosives works). Indoor sources include smoking, gas-fired appliances and oil stoves. Nitrogen oxides also arise from natural sources, such as volcanoes, lightning and bacterial action in soil. Globally, these sources contribute far more than man-made sources to total nitrogen oxides emissions, but the natural emissions are distributed so widely that their contribution to ground level concentrations is negligible in comparison.

Oxides of nitrogen may affect health directly. While nitric oxide is relatively harmless, nitrogen dioxide is a toxic gas, although even its milder effects do not occur at normal atmospheric concentrations. However, in the presence of sunlight and certain hydrocarbons, nitrogen dioxide can take part in the reactions which lead to the production of photochemical smog. Nitrogen oxides may also, together with sulphur dioxide, contribute to the acidity of rainfall.

The European Commission is now considering a draft directive which would limit nitrogen oxides emissions from stationary sources. Controls on emissions from mobile sources are progressively tightened under an EEC Directive series which applies to new vehicles.

Photochemical Oxidants

Photochemical oxidants are a group of chemicals found in the atmosphere of certain cities, e.g. Los Angeles in the United States. Their nature is not fully understood, but the group is known to include ozone, nitrogen dioxide and peroxyacyl nitrates — a group of organic oxidants, the commonest member of which is peroxyacetyl nitrate, conveniently abbreviated to PAN. These substances are formed in the air by the interaction of oxides of nitrogen and certain hydrocarbons. Strong sunlight is necessary to complete the reaction.

The basic substances which form the building blocks to the complicated reactions are exhausted to the air by vehicles, combustion and industrial processes.

The geographic features of the Los Angeles basin are such that it is singularly ill-favoured in relation to the production of photochemical smogs. Between May and October the smogs are noticed as a brownish-orange haze and build up to a sufficient concentration to cause eye irritation and coughing. Controls designed to reduce emissions from vehicles and other sources have been introduced in America in order to prevent this problem.

There is no evidence that photochemical smog is a problem in the United Kingdom, nor is there any suggestion that it is likely to become one in the forseeable future.

Carbon monoxide (CO)

Carbon monoxide is an invisible gas which is produced by the incomplete combustion of carbon-based fuels. The main source of man-made discharge to the atmosphere is from the exhaust of petrol engines. Concern has been identified regarding the concentration of CO in areas where there are high levels of traffic movement.

The effects of this gas on human health are well understood. It combines with blood in preference to oxygen. This means that anyone exposed to high levels of carbon monoxide will suffer from a form of suffocation, since the body is being starved of essential oxygen.

Measurements in urban areas have not shown levels which are likely to have a serious effect on health.

It has been determined that the working man or woman might be exposed to an average level of carbon monoxide of 50 ppm over an 8-hour period. High levels may occur in multi-storey car parks, underground garages, and other poorly ventilated areas affected by high numbers of vehicle movements. Even in these situations, it is unlikely that there will be any risk to passers-by, but special steps may be necessary to protect people working within such an area.

Deaths have been caused by badly maintained heating plant which have fumigated rooms, and by the deliberate inhalation of car exhausts. Another significant source of CO is cigarettes. Many smokers have much higher levels of CO compounds in their blood than non-smokers.

Description	Nonsmokers		Smokers	
	mean	range	mean	range
UK pregnant women	1.1		3.6	
Meat porters	1.6		5.1	
Office workers	1.3		6.2	
London office workers	1.12	0.1–2.7	5.5	2.2–13.0
29000 USA blood donors	1.39	0.4–6.9	5.57	0.8–11.9
3311 California longshoremen	1.3		5.9	
Munich population	2.36		7.38	
Rural Bavarians	1.03		6.06	

Table 1. Percentage carboxyhaemoglobin levels in smokers and non smokers *(Source: WHO, 1979. Environmental Health Criteria 13, Carbon Monoxide. (From Wakeham, 1976).)*

Carbon dioxide (CO_2)

The burning of fossil fuels (natural gas, coal and oil) is contributing to an increase in the amount of carbon dioxide in the atmosphere. It has been suggested that if this trend continues the surface of the earth will become slightly hotter than it is today. This is because carbon dioxide is better at transmitting short wave radiation from the sun than it is the longer wave infra-red radiation reflected back from the earth's surface. This is called the 'greenhouse effect'.

If, as some researchers predict, the level of carbon dioxide doubles in the atmosphere in the next 50 years, temperature patterns throughout the world would change. It could eventually lead to the melting of the polar ice caps, which would result in the sea level rising by several feet, producing widespread flooding.

Total suspended particulates

It seems that to every scientific theory, there is a counter-claim. Some researchers have discounted the importance of the 'greenhouse effect' by drawing attention to the quantity of particulates in the upper atmosphere. Their effect, it is said, is to reflect sunlight. The ultimate result, if emissions continue unchecked, will be the general cooling of the surface of the globe, and a mini ice-age. This has been called the 'blanket effect'.

Chlorofluorocarbons

Chlorofluorocarbon is the name given to a series of chemicals of varying numbers of carbon, hydrogen, chlorine and fluorine atoms; they are commonly referred to as "Freons" which is the trade name of an American manufacturer. They were developed in the USA in the early 1930s as a hazard-free refrigerant, but have found increasing application as the propellant (that is, the carrying substance) in aerosols. Concern has been expressed that these substances are now being emitted to the atmosphere in such quantities that they can cause significant changes to upper levels of the atmosphere. Research and mathematical modelling has not fully supported these claims, but the subject remains controversial. One alternative propellant is liquified petroleum gas; but while LPG is not regarded as a danger to the general atmosphere, it is highly inflammable and so may present more local dangers.

Asbestos

Asbestos has been identified as one of the more dangerous substances in general use. It is a naturally occurring mineral which exists in several forms, blue, brown and white being the most common. Ill-effects on health have been found when individual fibres become lodged in lung tissue. In this context, blue asbestos is generally accepted as being more dangerous than any other form. Asbestos has a variety of applications when heat resisting properties are required. It forms an important ingredient in the brake shoes of all forms of wheeled transport. Although it might be feared that large quantities of fibres could be released to the atmosphere whenever brakes are applied, this is not the case. Being a chemical compound, much of it is broken down by the heat of friction. The chemicals released do not create a hazard and the amount of asbestos fibre which survives to be released as dust from brake wear is extremely small.

Blue asbestos is no longer used in the United Kingdom. Factories processing asbestos are required to filter air before releasing it to the atmosphere. Special checks must be carried out to ensure that all asbestos waste is properly disposed of and that it cannot be discharged into the environment in a form which will endanger human or animal health. There is at present widespread public anxiety and uncertainty about asbestos in the home, in schools, hospitals and in workplaces where people may be incidentally exposed to health risks. Local authorities in many areas are trying to assess the degree of risk associated with the use of asbestos in buildings, particularly in council property. Environmental health officers share responsibility with the Health and Safety Executive for health and safety enforcement outside the factory, which includes supervision of the use of asbestos in new building work and its repair, maintenance and removal. County Councils in England and District Councils in Wales and Scotland are the responsible authorities in relation to the disposal of notifiable asbestos waste.

Heavy Metals

Lead is perhaps the most ubiquitous of the heavy metals. Certainly, its effects have been the subject of more detailed studies than have been devoted to any other element, and more data on environmental levels of lead have been collected than for any other single metal. It is possible that in future years another substance will be incriminated as being responsible for some as yet unidentified disorder. In the meantime, environmental workers are continuing to monitor a wide range of 'trace - elements' in the atmosphere: some 25-30 elements, mainly metallic, which although present in living tissues, were previously difficult to quantify. With advances in analytical techniques and understanding of the biological role of some metals, there is now more information on the accumulation of e.g. mercury and cadmium in the environment, and on the amount which reaches people, particularly in food.

Lead

In recent years, more concern has been generated about the effects of this heavy metal in the environment than about any other substance. Some aspects of lead pollution are discussed in the following paragraphs. Further information is available in the NSCA leaflet *'Lead and You — Reducing the Risks'*.

Lead in air arises primarily from motor car exhaust. Lead is added to petrol to improve its combustion properties. Whilst some is ultimately trapped in the vehicle's exhaust system, most is released to the atmosphere as fine particles. The dust may be inhaled, eaten as a contaminant of food, or simply transferred by fingers to mouths. Lead is also released into the atmosphere from lead works, which are subject to special controls to ensure that the emissions are minimised.

For some time pressure groups have been campaigning for the total removal of lead in petrol, since this is the most ubiquitous source of lead in the environment. In 1981, the Government announced a reduction in the present limit of 0.40 grams per litre, to 0.15 g/l by the end of 1985. The NSCA welcomed this decision as the most cost-effective interim measure, but in evidence to the Royal Commission on Environmental Pollution said that lead free petrol should be mandatory for all new cars at the earliest opportunity. When the Royal Commission's report on *'Lead in the Environment'* was published in April 1983, the Government immediately accepted its recommendation for lead free petrol. Negotiations are now underway within the EEC to amend the Directive which specifies the lead content of petrol, in effect to secure the removal of the present lower limit of 0.15 g/l. The NSCA also seeks a reduction in all sources of environmental lead. The Royal Commission made 29 specific recommendations to Government, designed to reduce human exposure to lead in air, dust, soil, food, paint and miscellaneous sources. The Government promised an early response on all points.

In some areas, where drinking water is acidic, there are likely to be elevated levels of lead in the supply of homes with lead water pipes/tanks. For this reason, Water Authorities treat drinking water to reduce the likelihood of it dissolving lead from supply pipes, and Home Improvement Grants might be available in some circumstances towards the cost of replacing old lead pipes and lead lined tanks.

Lead used to be a common ingredient in the pigments of many paints and old houses may have surfaces coated with many layers of such paint. Unfortunately, some children have developed lead poisoning by chewing these old painted surfaces, or, when the paint work is in poor condition, picking off and eating pieces of paint. Stripping woodwork is not necessarily an answer to the problem unless it is done in a manner which minimises the release of lead dust or fumes from the paint into the house.

Lead also occurs in some cosmetics (for example surma — an Asian form of eye shadow), in some hair dyes, and is used as an ingredient in some pottery glazes.

At one time, lead poisoning was not uncommon in people employed in the lead industry but this is no longer the case. There is, however, growing concern about the effects of lead on the health of young children. Recent research, notably in America, has shown that it is possible to demonstrate that children with high levels of lead in their blood do less well in intelligence and attainment tests than children with low levels of lead.

NOISE

Noise is on the increase in our society and it is recognised to be an unjustified interference with ordinary human comfort. One person can cause as much disturbance to his neighbour as can a factory. The thoughtless use of a radio, an unsilenced motorbike or a badly maintained lawn mower can be just as annoying as a large piece of industrial plant. It is, therefore, up to us all to minimise the amount of noise which we produce — that is, if we care for our local environment.

Sound

Noise is unwanted sound. The science which includes the study of sound, acoustics, is relatively complex and cannot be fully explained here. For a detailed explanation, see the NSCA's booklet 'Noise and Society'.

Sound is caused by very small changes in air pressure. The manner in which sound spreads is rather like the way in which ripples travel out over the surface of a pond after a stone has been thrown into the water. The idea of a ripple or a wave can also be used in trying to understand some of the concepts associated with sound measurement. The difference in height between the top of a wave and the average level is known as the *sound pressure*. The number of waves which pass a fixed point within a second is called the *frequency*. The distance between successive waves is known as the *wavelength.*

Almost all sounds which we hear contain a mixture of frequencies: the higher the tone, the higher the frequency.

Measurement

The range of sound pressures in everyday life covers a spectrum from the threshold of hearing to values over a million times higher. To cover this range with manageable numbers a logarithmic system of measurement has been devised. Its basic unit is the decibel.

The human ear does not respond to changes in sound pressure in the same way as electrical measuring equipment. The ear is relatively inefficient in its assessment of very low or very high frequency noise. An electronic circuit has therefore been devised which enables measurement of sound to be made so as to take less account of the lower and higher frequency components. This circuit is called the A network and is nearly always used if noise is being studied in relation to human response or exposure. Table 2 shows some typical sound levels.

Sound pressure in bar	Sound level in dB(A)	Environmental conditions
	140	
1 mbar	134 130	Threshold of pain
	120	
100 ubar	114 110	Loud discotheque
	100	
10 ubar	94 90	Unsilenced pneumatic drill at 5 m
	80	Inside motor bus
1 ubar	74 70	Average traffic on street corner Vacuum cleaner at 3 m. Conversational speech
	60	
0.1 ubar	54 50	Typical business office
	40	Living room, suburban area Refrigerator humming at 2 m
0.01 ubar	34 30	Library
	20	Bedroom at night
0.001 ubar	14 10	Broadcasting studio
0.0002 ubar	0	Threshold of hearing

Table 2. Some commonly encountered noise levels (dB(A)). *(After Jens Broch.)*

Noise and Nuisance

Much of the noise which irritates the public is totally unnecessary, whether of domestic or industrial origin. It is possible for anyone to enjoy music either in their home or in a public place at any level of loudness which they can tolerate, without disturbing others. All that is required is a pair of headphones (the user should, however show some moderation in volume or permanent damage can be done to hearing.) Industrial noise can be controlled by the use of "quiet processes" and buildings adequate to contain the sound.

Anyone living in the United Kingdom has a legal right to be protected against noise nuisance. The most usual procedure is to complain to the environmental health department of the local authority. Investigations will be carried out and the situation assessed impartially. If it is felt that a nuisance is being caused, the person responsible will be required to stop or minimise the disturbance. Generally, advice will be given on how the necessary reduction in sound levels can be achieved. This may involve the modification of machinery or the re-design of part of the building. In the case, for example, of nuisance being caused by a pop group practising in a house, it may be necessary for an alternative rehearsal room to be found.

There is an automatic right of appeal against this form of action. It is the court and not the official who has the final say. It is also possible for individuals to take their own legal action against excessive noise if they do not wish to involve the local authority.

Noise from Loudspeakers in the Street

It would be impossible to ban all sources of noise. In certain cases noise is produced to warn the public. An example of this is the siren of a police car, ambulance or fire engine.

Amplified music (or at least sound) is used by ice cream salesmen. This practice is accepted nationally but chimes must not be sounded between the hours of seven in the evening and noon the following day. A code of practice has been drawn up to give further guidance to ice-cream salesmen. Other sources of noise subject to specific control are burglar alarms and model aircraft.

Construction Sites

Building and demolition are, by their very nature, noise-producing activities. The equipment used and the techniques involved can generate very high levels of sound. Control is often difficult because of the temporary nature of operations.

Legislation exists which enables an environmental health officer to set maximum noise levels from such sites — the point of measurement is usually the nearest noise sensitive property. Hours of work can be limited and particular methods of operation specified; e.g. the use of the falling weight or power hammer in pile driving can be replaced by other techniques which vibrate or force the pile into the ground.

Pneumatic breakers, the tools used to cut through the road surface prior to excavation, can be fitted with silencers. (This is rather an inaccurate description as they are not actually silenced but merely quietened.) The steel point can be treated to prevent it "ringing" by fitting a rubber sleeve at its mid point. Compressors can be so built as to contain most of the engine noise within the weather-proof body. Such steps enable roadworks to be carried out without annoying the surrounding community. But all too often those responsible for such works neglect this simple aspect of their duty.

Noise Abatement

Generally noise control is not solely limited to legal action to stop nuisance. Powers were granted to local authorities in the Control of Pollution Act 1974 to take long term steps against industrial noise. The idea is to stop the general upward creep in the noise level, within an area designated a "noise abatement zone".

This action can only be taken if the local council feels that it has adequate staff to carry out the necessary procedures. The basic approach is to define an area, determine the location of industrial and commercial premises and measure external noise levels at numerous locations along the boundaries. The measured levels are recorded in a register. If an industry increases its noise level an offence will be created unless the local authority can be convinced that a new set of conditions should be imposed. It is also possible for action to be taken to reduce noise levels from premises within a noise abatement zone even if an existing noise is not causing a nuisance.

Sound level meter in use, inside the workplace (*Photograph courtesy of Bruel & Kjaer (UK) Ltd.)*

It is very time consuming to collect all the data necessary to establish and monitor an area. Many local authorities feel that this exercise is not as important as their other duties so there are only a few of these areas throughout the entire country. An enquiry to the environmental health department of the local authority will enable anyone to find out if there is a noise abatement zone in the neighbourhood. If there is, it should be possible to see a map of the exact location of the area and the register of recorded noise levels.

Noise and Planning

Whilst it is nearly always possible to reduce noise once it has occurred, it is better to prevent nuisance initially; this applies particularly to the case of noise from industry and commerce. Before any building can be erected or extended, planning permission must be obtained. One aspect of any new development which is given careful consideration is its noise potential. Some processes generate so much sound that it would be reasonable to prevent their introduction into any residential area. In other cases modifications can be made to reduce the impact of a factory on its neighbours.

An example of noise which is difficult to control is that from mineral extractions. Coal, gravel, limestone and similar materials are only found in a limited number of locations. Geological problems reduce further the total number of sites which can be commercially worked. It is therefore inevitable that permission will sometimes be given for a quarry or open cast coal site on land close to housing. Where this happens complaints are frequently made about matters which cannot be remedied. Noise from blasting, for example, could be restricted but no one has yet invented silent explosives.

Similarly, it is general policy to take heavy road traffic away from the centre of small towns by constructing by-passes. This means that the town dwellers enjoy more quiet whilst some people living in rural peace experience disturbance when a new road is carved through the fields. There is no simple solution to this sort of problem.

The need to avoid building "bad neighbour" factories, airports or new roads near to existing houses has been clearly understood. It is equally important not to build new houses near to an existing noise producer.

Many local papers carry weekly lists of planning applications. This gives environmental groups and other interested people the chance to monitor what is going on and to have a say in development proposals.

Road Traffic Noise

The most widespread source of noise nuisance is road traffic. Unfortunately, there is little that can be done to achieve great reductions in existing levels.

The amount of noise which any new vehicle may produce is limited. Once on the road, the use of faulty silencers is illegal and vehicles must not be driven with 'excessive noise' but the problem of enforcement is so great that only the most flagrant offenders seem to be caught. This may in part be due to the attitude of the public. There is a general impression that noise is related to power and that the more "throaty" a motor cycle or car, the better its performance.

As previously discussed, taking roads out of the centre of population often transfers noise to a new, although perhaps a smaller, group of people. If concrete barriers 20 metres high were to be built on either side of the new road the only people to hear the traffic would be the dwellers of high rise flats. This solution would be inordinately expensive, ugly and unacceptable to drivers and passengers alike. Noise insulation can be provided for houses by the installation of double glazing and ventilators with built-in silencers. Whilst such steps usually reduce internal sound levels they deny the occupiers the opportunity to sleep with their windows open. Furthermore, nothing can be done to protect the garden or other recreational areas near busy roads.

New buildings can be designed so that they face away from major highways. The use of windows on the road side can be minimised so as to reduce noise penetration. In acoustic terms this may be an acceptable solution. Architec-turally, such a building sometimes has a prison-like appearance.

It is often worthwhile looking along the route of new roads to see exactly what steps have been taken to deal with the problem of noise. It is also interest-ing to try to draw a proposed route for a by-pass, link road or other improve-ment road on a map.

This will highlight the problems of minimising the length of the route, avoiding houses, leaving farms intact, etc.

Aircraft Noise

Many people who live near to airports find noise from aircraft annoying. Limited restrictions may be imposed on the hours during which aircraft may operate from the airport and the way in which they take-off. But such action can never be sufficient to prevent disturbance. Aircraft engines are by their very nature noisy, especially jet engines, and although work is carried out to quieten engines and reduce sound levels, for many people the problem remains acute. As in the case of road traffic some comfort can be given by improving the noise insulation of those homes nearest to the approach and take-off areas of the air-port. Such schemes do not exist as a right, being drawn up by the operators of certain airports. In general, insulation schemes are also limited to civil aircraft — for some reason military aircraft are not thought to be so difficult to live with.

At the time when this booklet was being written considerable debate was being given to the location of the third airport for London. Heathrow and Gatwick are well developed and it is thought that additional facilities should be provided elsewhere. The difficulty of finding a suitable site, within a reasonable distance from the capital, has taxed civil servants and politicians alike. One possible location could be Stanstead but currently no decision has been made public on this matter.

Hearing Loss

It is generally accepted that exposure to high levels of noise can damage hear-ing. The risk of this happening is related to the level of noise and the duration of exposure. If a noise level is so high that it is necessary to shout to someone a metre away, and if exposure to such noise lasts for about eight hours a day, during normal working hours, there is a real danger to hearing.

Higher noise levels over shorter periods produce similar effects. The quality of sound, be it a machine or music, is not relevant to the effect. People who are exposed to high levels of noise both at work and at leisure are most at risk. As mentioned earlier, damage can even be caused by highly amplified sound heard through head phones.

If equipment in a factory is found to produce excessive noise means should be found to change the machine or the process. Alternatively, it may be possible to provide the operator with a quiet area in which to work. Only as a last resort should ear protection be regarded as an appropriate remedial action.

Regardless of the very real nature of this problem it seems most unlikely that discotheques will display notices at their entrances saying, "Loud noise can damage your hearing!" It is even less likely that patrons will be seen wearing hearing defenders.

WATER POLLUTION

Water is considered to be polluted when it is altered in composition so that it becomes less suitable for any use to which it could have been put in its natural state. Such pollution is often caused by uncontrolled disposal of sewage and other liquid wastes (whether of domestic, industrial or agricultural origin). Over-fertilization of land also causes problems when materials used are washed into streams and rivers.

Water Cycle

Water takes part in an endless cycle of movement. It is evaporated from sea and land alike, eventually condensing into the individual droplets which make up clouds. The next step in this process is precipitation — the production of rain, snow or hail. Some of the water which falls on land will run off again, eventually reaching the sea by streams and rivers. The remainder will be taken up by the soil, some soaking into the groundwater reservoirs or being taken up by plants and animals whilst the rest is evaporated again. An illustration of this cycle is shown in Figure 4.

Background

At one time all of the rivers in Great Britain were clear and supported abundant fish and animal life. As cities began to grow in the 18th and 19th centuries the situation changed. Although the invention of the water closet did much to improve domestic sanitation it did little for the general environment. Open sewers ran along many streets discharging their filth into the nearest river. Eventually the concentrations of sewage became so high that most aquatic life was killed. Rivers became stinking stretches of highly polluted water. For a few days in 1858 Parliament could not sit because of the stench from the Thames.

In addition to killing fish there was another, greater threat to the local population. Many rivers were also used as sources of drinking water. There is a real risk to health if water used for food preparation and drinking purposes is contaminated by sewage.

Recognition of the problem led to new laws and to improvements. Methods of treating waste water were developed so that the most offensive matter was removed before it could be discharged to any water course. The purification of drinking water became a national practice. Some rivers are still far from satisfactory but water from the tap should always be fit to drink (but note the earlier section on Lead.)

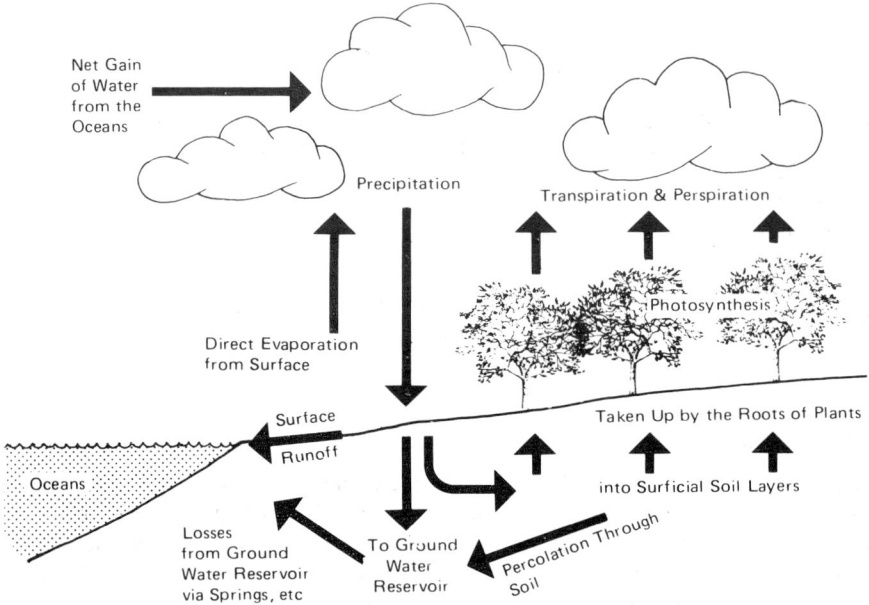

Fig. 4 The Natural Water Cycle.

Drinking Water

Untreated groundwater is normally much safer to drink than untreated surface water because the ground acts as a filtering medium. No potential supply can be regarded as absolutely safe because of the danger of pollution from refuse tips, domestic or industrial waste waters or major spillages of toxic chemicals.

A considerable amount of the water we use is taken from rivers. Treatment is involved which, in essence, consists of removing the floating debris, settling out suspended solids, and filtration and chlorination prior to storage and use. It should be possible to obtain details of the exact nature of treatment used locally from the water authority.

Sewage Treatment

The need to treat used water before returning it to the environment has been mentioned above. Most buildings are provided with drains which connect to sewers leading finally to the sewage works. Some isolated houses cannot be connected to the main drainage system so watertight tanks are provided which receive all liquid waste. In some cases it is possible to provide treatment facilities at remote houses but this is uncommon.

Sewage treatment begins with the removal of rags, wood, dead dogs and all the other large objects which find their way into the drains and could damage machinery. Grit and sand are allowed to settle out. Solid matter is removed from the water by a process of separation.

The relatively clear liquid is passed to special processing beds where it is trickled into tanks of clinker or gravel. Here, bacteria and oxygen work together to digest the remaining waste matter leaving a clear effluent to be returned to the river.

The solids previously separated out still carry with them a considerable quantity of water. This thin mud-like liquid is called crude sludge. It can be further processed before being disposed of but there is increasing use of this material as a form of agricultural fertilizer.

There are dangers associated with the application of sewage sludge to land. If the material contains organisms harmful to man it is possible that drinking water supplies could be contaminated. Various organisms occur in sewage which can affect cattle, for example human tapeworm eggs. Stock must be excluded from fields where sludge has been applied for a sufficient period to ensure that there is no risk of infection. This is very important if selling unpasteurised milk.

The operation of a sewage works is relatively straightforward. It is sometimes possible to arrange with the water authority for a conducted tour to be given to an interested group.

Water Pollution Control

In general it is possible to treat domestic sewage so as to produce acceptable water. This is because the waste is largely organic. Most industrial waste is inorganic and is far more difficult to process.

In some cases, wastes are so toxic that they cannot be discharged to the sewer and have to be removed by special tankers to expert disposal companies. In others, firms have to install their own treatment plant. Either method is expensive and an unscrupulous firm may be tempted to cut costs by running liquid to the drain at night. Pollution control officers do not always work the expected nine-to-five day and by clever detective work can sometimes trace such substances back to their source.

It is not only the irresponsible industrialist who causes pollution; just as culpable is the motorist who empties used sump oil down the drain.

River Management

Proper operation of sewage works and vigilant enforcement of pollution control has led to a real improvement in some of our rivers. The Thames, which once degenerated into a massive open sewer, is gradually returning to its original condition. The varieties of fish which can breed there are increasing; there is even action in hand to bring salmon back to the river.

In other parts of the country, the situation is not so reassuring. It will be many years before some rivers can even be considered fit for water sports.

Coastal Pollution

Two important aspects of coastal pollution can be identified which may affect large numbers of people.

The first is oil — whether it be from a major tanker disaster or the waste discharged over the side of a ship. The black tar-like oil sometimes found on beaches can be a nuisance to holiday makers but spells death to sea birds. Little more can be done to reduce the risk of ship-wreck, so it seems inevitable that there will be an occasional catastrophe when an oil tanker breaks up near the shore. Enforcement of laws concerning dumping oil and waste into the sea is difficult. Responsibility must rest with the captain of each vessel and some seem more willing than others to guard the environment. Once oil is in the sea, and the ship has steamed on, it is difficult to prove that an offence has been committed.

Oiled guillemot between rocks *(Photograph courtesy of the RSPB and Michael W. Richards, Photographer.)*

However, response to incidents could be more speedy. When oil is found washing ashore, at present it is the local council's responsibility to clean the beach, while local volunteers try to save the lives of oiled-up sea birds. Unless the culprit can be identified costs have to be borne locally; yet the problem seems one which requires the resources of central government.

The second aspect of pollution of the sea which merits special mention is the practice, in some seaside resorts, of piping crude sewage into the sea. Often this sewage is not even given the most basic treatment and an unfortunate bather can find himself swimming amongst all sorts of unmentionable material.

Standards have been laid down which relate to the quality of bathing waters but they only apply at major resorts. Public demand may eventually lead to a more general improvement.

WASTE

The insanitary collection and disposal of solid waste creates serious health hazards: it can encourage the breeding of flies, rodents and other carriers of disease. Such operations can cause serious public nuisance and contribute significantly to the deterioration of the environment, increasing water, air and soil pollution.

Refuse disposal involves three separate features; the collection and transport of waste; processing by means of incineration, composting, compaction and removal of recycleable waste; and final disposal of waste by controlled tipping.

Methods of Collection and Disposal

Storage of refuse can be a local problem which is at its most serious in areas of dense housing and in some flats and tenements where inadequate provision has been made for rubbish storage.

It is not practicable for individuals to make their own disposal arrangements. Small incinerators seem incapable of operating without smoke whilst very few gardens could be used as rubbish dumps without causing distress to the neighbours.

Disposal remains difficult because of the potential nuisance involved. At one time many local authorities employed men to hand-sort all refuse so that material with commercial value could be recycled. It is an exceptionally unpleasant occupation, especially during the warmer period of the year when rubbish decomposes quickly.

In order to reduce the bulk of material to be disposed attempts have been made in some areas to encourage people to separate out paper, glass and metal (see also the section on Recycling). Generally, however, all material is placed in a dustbin and ultimately disposed of as waste — the one possible exception being iron and steel which can be lifted from a conveyor belt by means of a powerful magnet during initial handling operations.

Until recently there was a trend towards the total incineration of all refuse — in some cases the resultant heat could be used for a variety of applications. The costs of building and operating incinerators have become so great that they have discouraged this practice. Another factor which adds to the cost of burning rubbish is the need to prevent air pollution both from smoke and toxic elements.

Various methods of grinding, handling and sorting domestic refuse have enabled compost-like materials to be produced. There have, however, been doubts about the application of this substance to land because of possible risk of contamination from poisonous materials. Furthermore, the energy involved in this form of process is considerable, especially if the cost cannot be recovered by sale of the marketable commodity.

Unwanted household rubbish dumped on land *(Photograph courtesy of Richard Hawkins.)*

In some ways the clock seems to have been turned back and waste disposal authorities throughout the country are returning to controlled tipping. This operation is not energy intensive. A suitable depression in the ground is found — geological investigations are carried out to minimise the risk of pollution of groundwater — and tipping begins. Domestic and commercial refuse are deposited in layers some 3 metres thick and covered with soil or similar inert material. This minimises the opportunity for flies, birds and rodents to feed and breed on the site.

Gradually the area will be filled with layer upon layer of rubbish. Under normal conditions it will slowly rot away, settling slightly in the process. Methane is often produced by the bacteria responsible for the breakdown of materials. The possibility of removing this gas, which can be used as an industrial fuel, is currently under investigation at various waste disposal sites.

Building upon the tip cannot normally take place for some years after the last waste is dumped. This is because of the unequal settlement of the land and the risk of gases seeping into any structure built upon the site. In the short term the area can be landscaped and used for public open space but eventually the site which had previously been rendered unsuitable for developments because of the presence of the large hole can be used for almost any purpose.

Special arrangements have to be made for the effective and safe disposal of certain types of material such as hospital waste; medicine and poison; and bulky household discards, e.g. old furniture. It is amazing that some people find it easier to drive into the country and dump an unwanted mattress in a layby than to take it to a proper waste disposal site.

Industrial Waste

Certain chemical and industrial waste cannot be disposed of by uncontrolled burial in general tips. In some cases deep burial can be specially arranged providing that the public authorities give their consent. Systems of notification operate and records must be kept to ensure that material will not be accidentally excavated at a later date.

Some substances never break down and have to be disposed of in special ways. Sometimes incinerators can be built and designed to ensure complete destruction of difficult chemicals. There are materials which cannot be burnt, for example radioactive waste.

Deep burial or disposal in deep parts of the sea bed are practised.

Recycling

In order to conserve valuable raw materials, some components of domestic waste can be separately stored rather than thrown into the dustbin. Bottles, food and drink cans, paper, plastic, metal and fabric all have commerical value. Increasingly the public seems willing to use bottle banks, so enabling otherwise unwanted glass to be returned to the glassworks. Aluminium drink cans, although bulky, can be returned for remelting. Other food cans, if washed, can first have the valuable tin plate removed then be melted down so as to recover the steel. There is even a market for used car oil which, once reprocessed, can be used as a fuel (some scientists do have reservations about this because of the possibility of oil being contaminated by lead from the petrol).

The trend towards recycling is growing and will become increasingly common as finite raw materials become less freely available.

LAND POLLUTION

Introduction

Soil pollution is usually a result of poor agricultural practice, incorrect waste disposal or even fallout from atmospheric pollution. Soil is becoming increasingly polluted with chemicals, including heavy metals which can reach the food chain, ground and surface water and ultimately be ingested by man.

Solid Wastes

Waste disposal has been discussed in a previous section. Its mention here serves to highlight the internal relationship between different sectors of the environment.

Industrial Activity

Long term land contamination can be created by certain industrial uses even if there is no direct intention to dispose of large quantities of toxic material within a site. Examples include scrap metal yards and sites occupied by plant previously used for the manufacture of gas from coal. Land pollution is also caused by the dumping of large amounts of waste materials produced from the mining of coal and minerals and from the smelting of metals.

In each of the above cases toxic or harmful substances can leach out and find their way into the soil. It is therefore necessary for expensive and comprehensive investigations to be carried out before such sites can be redeveloped because of the possible risks of contamination.

Agricultural Activity

Fertilizers intended to improve soil can also cause contamination from impurities. Herbicides, insecticides, fungicides and soil conditioners have been applied to some areas without sufficient regard being given to their long term effect. Such substances can remain unchanged, in the soil, for long periods of time. Information on the way in which some of these materials break down also appears sketchy.

In the widest context of land pollution, concern must be directed towards the indiscriminate use of a wide range of agricultural chemicals. This issue was clearly and eloquently discussed by Rachel Carson in her book *"Silent Spring"* (Penguin). This highlighted the threat to wildlife in general from the careless use of pesticides and similar substances. No writer since has so brilliantly reported on this subject.

ENERGY

Conventional Sources

All conventional fuels which are available in significant quantities were produced millions of years ago. Coal, oil and natural gas owe their origins to massive deposits of vegetable matter during the carboniferous period.

Subsequently, the material has altered in nature to form solid, liquid or gaseous fuels.

The first fuel to be used to any great extent was coal. Perhaps one reason for its decline in popularity was the difficulty of achieving automatic control which is as attractive to the industrialist as to the householder.

Although Great Britain has reserves of oil and gas, their total potential is relatively short compared to that of our coal reserves. Even coal deposits, however, are of limited size and may not last for more than three hundred years.

About one half of the coal and one third of the heavy fuel oil used in Great Britain is burnt by the Central Electricity Generating Board. They have stated that not only must plans be made for a time when coal ceases to be available but that consideration must be given to the wisdom of burning coal at all. Being rich in a wide range of organic substances, there is clear logic in regarding coal as a raw material for the heavy chemical industry.

THERMAL POLLUTION

Processes where energy is converted from one form to another tend to be inefficient. The generation of electricity is a case in point. Conventional modern power stations only convert some 35% of the total heat in the fuel to electricity. About 10% goes up the chimney, leaving 55% in cooling water. At one time this warm water was returned to the local river but this did not seem very beneficial to aquatic life. The next step was to achieve the necessary cooling by the use of tall towers. These are dominant features of the building complex of most power stations. Water vapour can often be seen condensing into a dense white cloud at the top of a cooling tower. Under certain weather conditions condensation occurs at a greater altitude so that the location of a power station is clearly marked by the clouds forming in the sky almost directly above.

In an attempt to minimise waste, various schemes have been undertaken by the

Central Electricity Generating Board. One idea has been an eel farm — these unattractive creatures flourish in warm water. Another has been a 20 acre greenhouse at Drax power station which is being used to produce tomatoes.

Further ideas which can make use of water at about 30° Celsius are under review, although other industrial processes do not tend to produce such significant quantities of waste heat. Rising energy costs, however, have caused all manufacturers to examine their operations with a view to increasing energy efficiency.

Nuclear Power

Britain embarked on a programme of investment in this form of electricity generation in 1955. Eight Magnox stations are now operational and five Advanced Gas-cooled Reactors (AGRs) are all in advanced states of commissioning. At the time of writing, consideration is being given to the construction of a Pressurised Water Reactor at Sizewell.

Considerable technical information on the design of various types of reactor referred to in the previous paragraph can be obtained from the CEGB as can their case for the generation of power by nuclear means. Other organisations will present counter arguments. It is not within the scope of this booklet to enter into the nuclear debate. While there has never been a major emission of radioactive material from a nuclear power station, there have been alarming incidents, notably in the USA.

As for the UK nuclear fuel processing industry, various incidents have been reported at the site formerly known as Windscale.

Public concern has also grown over recent years in relation to the policy of transporting radioactive waste by rail from nuclear power stations to processing plant. The containers used for this purpose are massive and said to comply with all relevant safety standards, but some testing has only been carried out on scale models, which is not satisfactory to all sides.

Alternative Energy Sources

Interest has developed in the possibility of extracting energy from natural sources. Various options have been considered but all are currently of limited application.

Solar Energy

The generation of electricity for commercial purposes from solar cells is not practicable at present due to the cost of cells compared with their electrical output. The reliability of solar furnaces or boilers has not yet been established. Domestic hot water from various forms of solar collectors is feasible in some circumstances. Unfortunately, all these systems have a limited application in the UK because of the latitude and the lack of constant sunshine.

Wind Power

Whilst the use of windmills has been established in England for corn grinding since the end of the 12th century, application to 20th century technology has not fully developed. Small mechanical windmills may be of limited value for water pumping in isolated locations but the windmill is unlikely to become a major source of electricity in national terms. Just like the sun, the wind cannot be guaranteed to be there at the right time or with sufficient power to ensure adequate generation. Modern windmills are unsightly because of the size necessary for efficient operation. Noise is also likely to be a problem. Countryside tranquility would be submerged under a barrage of noise from the spinning blades.

North Leverton Mill with West Burton Power Station in background *(Photograph courtesy of the Central Electricity Generating Board.)*

Hydro Electric Power

Hydro electric power is produced from fast flowing water. Few sites in Britain have sufficient head or sufficient flow, thus the scope of this technique is limited.

Wave Power

As yet no equipment has been developed which, when scaled up, will be available to generate electricity at commercial rates. Further research is in hand. Wave power, at least in Britain, has a seasonal peak throughout the winter which almost matches electrical demand.

A variation on this theme is tidal power. The tidal flow of water into and out from a natural estuary can be tapped by the construction of a barrage across the estuary mouth. This is feasible but extremely costly. Furthermore, the ecological consequences within the area of contained water could be catastrophic.

Geothermal Power

This is the power obtained from the heat of the earth. In areas subjected to "recent" volcanic activity, such as New Zealand, Iceland and parts of America, steam can be produced, allowing electricity to be generated. In Britain, temperatures are much lower, providing only a potential for hot water.

Heat Pumps

The principles of a heat pump are demonstrated in the domestic refrigerator. Heat is absorbed from one point (the ice-box) and released at another (the exchange at the back). Similar equipment can be used to extract heat from a relatively cool source, e.g. a stream, and use it for providing hot water for domestic supply. The scope is very limited. Costs are attractive because less energy is required to transfer heat from one medium to another than to provide equivalent heat to the medium in which the temperature is being raised.

THE FUTURE

According to opinion polls in several countries, improvement of the environment is considered a priority by the public. Pollution abatement remains important to most governments, if only in recognition of the electorate's concern. Emphasis is, however, being switched towards the conservation of resources and improvement of the quality of life. There is a trend towards countries acting together in relation to major environmental issues. High on the list of action is for developed countries to reduce their energy dependency and help the Third World to develop its own resources. Countries must co-operate in the development of harmonious world evolution and work together even when divergent short-term interests, such as industrial competition, intrude. They should collect and study data on world environmental trends on an international basis because pollution recognises no boundaries.

Environmental policies must now be viewed on a global scale. In 1972 the United Nations Conference on the Human Environment at its Stockholm Meeting agreed a number of principles to guide the peoples of the world in the preservation and enhancement of the human environment. Included were the following:

"• The natural resources of the earth including the air, water, land, flora and fauna and especially representative samples of natural ecosystems must be safeguarded for the benefit of present and future generations through careful planning or management, as appropriate.

• The capacity of the earth to produce vital renewable resources must be maintained and, wherever practicable, restored or improved.

• Man has a special responsibility to safeguard and wisely manage the heritage of wildlife and its habitats which are now gravely imperilled by a combination of adverse factors. Nature conservation including wildlife must therefore receive importance in planning for economic development.

• The non-renewable resources of the earth must be employed in such a way as to guard against the danger of their future exhaustion and to ensure that benefits from such employment are shared by all mankind.

• The discharge of toxic substances or of other substances and the release of heat, in such quantities or concentrations as to exceed the capacity of the environment to render them harmless, must be halted in order to ensure that serious or irreversible damage is not inflicted upon ecosystems. The just struggle of the peoples of all countries against pollution should be supported.

• States shall take all possible steps to prevent pollution of the seas by substances that are liable to create hazards to human health, to harm living resources and marine life, to damage amenities or to interfere with other legitimate uses of the sea.

• Economic and social development is essential for ensuring a favourable living and working environment for man and for creating conditions on earth that are necessary for the improvement of the quality of life."

These principles remain valid foundations on which to continue to build a programme of environmental protection. Progress towards meeting the objectives behind these principles will depend on the commitment of individuals, action groups, governments and nations. The living conditions of future generations is dependent on our actions today and in the immediate future.

ABOUT THE AUTHOR

Michael J. Gittins is an environmental health officer who has developed a great interest in all aspects of pollution control. He is a chartered engineer, a Member of the Institute of Energy and a Member of the Institute of Acoustics. He has presented various papers at national and local conferences on a range of pollution issues including chimney height design, noise control from cooling towers, measurement of air pollution, environmental lead and the control of noise from construction sites. For some years he has produced a regular column on pollution topics for the Institution of Environmental Health Officers and is now providing occasional contributions and book reviews for other journals. He also serves on British Standard committees on vibration and incinerator design.

Printed and published in England by the National Society for Clean Air
136 North Street • Brighton • East Sussex   ISBN 0 903474 18 2